Exxxtra Special
A Miracle Named Halo

Written by Kafi Hunter | Inspired by Halo Victoria Ray
Illustrated by Cameron Jones

Exxxtra Special: A Miracle Named Halo

Text copyright © 2024 by Kafi Hunter
Illustrations copyright © 2024 by Cameron Jones
Edited by Pen to Print Editing & Publishing

All rights reserved. No part of this publication may be reproduced, distributed, or transmitted in any form or by any means, including photocopying, recording, or other electronic or mechanical methods, without the prior written permission of the publisher.

Revolt Renaissance Publishing

Library of Congress Control Number: 2023919794

Dedication

To my daughter Halo Victoria Ray, my masterpiece from heaven, thank you for choosing me to be your mother. Your light shines bright and changes the atmosphere. You are a promise kept by God and I am eternally grateful!

Love, Mommy

"Mommy, tell me a story!"

Halo's smile was a hundred rainbows.

"What kind of story do you want to hear baby?"

Mommy responded.

"I want to hear a story about a superhero!"

Halo exclaimed.

"Well, that's easy."

Mommy paused and ushered Halo to her chair.

"Have I ever told you about the little light that grew inside of Mommy's belly?"

"Fireflies in your tummy, Mommy?"

Halo rubbed her own belly and was confused because she didn't understand how Mommy could have a light growing inside of her.

"Everybody was so excited about the little light that was growing inside of me." Halo blinked her eyes in curiosity.

"Mommy and Daddy were ecstatic."

"What's 'ecstatic,' Mommy?"

"Extra excited," Mommy responded.

"As a matter of fact, Daddy and I went to the doctor one day and they told us the light in my belly was going to be extra special?"

"How was your light going to be special, Mommy?"

Halo showed her wonder with each question.

"The little light was going to be oh so special," Mommy started, "because the little light was going to be a little girl with an extra 'X' chromosome." Mommy continued her story.

"Mommy and Daddy were scared at first, but then we both got ourselves a cup of courage."

"Did that make you and Daddy brave?" Halo asked.

"Oh, so brave!" replied Mommy. "We decided to embrace our extra special light and come up with baby names,"

"I wanted to name the little light Harper Grace."

"Daddy wanted 'Victoria' for the middle name, after Grammy."

"Then, one night, the Spirit told me to name the little light Halo."

"Wait!" Halo ecstatically exclaimed, "That's me Mommy! Halo Victoria Ray!"

"So I'm the little light?" Halo jumped up and down with a smile that was magical.

"Oh, but you were ready to shine so bright!" Mommy confirmed.

"You were so ready to shine! You wanted to get here soooo soon! Oh my goodness! You came extra early!" Mommy took a breath, then continued.

"You didn't want to wait nine months, or ten." Mommy explained.

"When did I want to get here Mommy?" Halo questioned.

"After just six months!" Mommy paused and remembered the very moment Halo was born.

"But, April 16th, 2017, you were coming! Nothing was going to hold you back from getting here!" Halo giggled sweetly.

And that day was ohhhh so special, my beautiful light. Do you know why?"
"Because that's the day I came out of your belly, Mommy." Halo responded matter-of-factly. "Well, that's right," Mommy smiled. "But it was also Easter Sunday and you were showing us that miracles still exist."

"Annnnnnd," Mommy continued, "April 16th is the birthday of my Aunt Jackie. She's passed away now, but I just knew we had our very own angel looking out for you!"

"And Mommy sure needed an angel because you were my fifth pregnancy."

"Well, that's refreshing." Halo responded.

She was happy Mommy was able to have her very own angel.

"It sure was," Mommy started. "You came out so small! Only one point five ounces... tiny, tiny baby! The doctor said you only had a sixty percent chance of living, but you weren't having it!"

"You had to stay in special care, called nicu, for five months."

"Did that make you and Daddy scared Mommy?"

"It did," Mommy considered. "But your light never dimmed."

"You frightened us again one time because you almost coded because you were pulling the breathing tubes out of your nose."

Halo's eyes got real big as Mommy continued.

"It was like you were saying, 'I got it Mommy, I'm going to be able to breathe on my own!'"

"Well, look at me now Mommy!" Halo glowed.

"My little light," Mommy smiled and pinched Halo's cheeks.

"You know what?" Mommy thought. "You were actually named the 'Princess of the nicu.'"

"A princess. Mommy?!" Halo illuminated.

"Oh my gosh, yes!" Mommy started.

"You were such a fighter that UAMS put you on their 'Wall of Hope' for everyone to see!" "Wow!" Halo exclaimed.

"The doctor didn't know if you would be able to speak very well, and they thought you might be blind."

Halo didn't even have a chance to respond before Mommy continued.

"But *baby*! You speak so well, so fluent and articulate, you may be a debater one day."

Halo didn't know what a debater was yet, but Mommy's voice was funny when she said 'baby."

"And your sight goes past your eyes." Mommy explained.

"You see so well, many people call it prophecy. Both your grandma's say when they're down they want to see Halo."

Halo thought about all the praise Mommy was giving her....

"Well sometimes I get scared and have to take a cup of courage too Mommy."

"What scares you, baby?" Mommy asked.

"Tall buildings and climbing stairs by myself." Halo responded as she felt herself shrinking. Mommy considered Halo's words and rubbed her back in assurance.

"I know you're going to continue to conquer because you're a star!" Mommy encouraged.

Mommy stood up and spun Halo around in a circle.

"Halo The Boss!" Mommy roared as she pulled out two magazines and pointed to pictures of her little light, Halo.

"Well," Halo began as she hugged Mommy's waist, "You've always had my back, Mommy."

"And you always will be my little light."

Mommy squeezed Halo tight in a hug that decorated the whole entire world with hope and cheer.

About The Author

I am a mom that almost died twice giving birth due to severe preeclampsia and HELLP Syndrome. The first time was on November 5th, 2011, at Baptist Health in Little Rock, Arkansas. I had to deliver my child at 22 weeks and 6 days old naturally. Haley Vaughn Jones my angel daughter was a live birth, but she did not survive past two hours because she was severely underweight, and her lungs were not developed. The second time was on Resurrection/Easter Sunday April 16th, 2017, at UAMS in Little Rock, Arkansas. I delivered my child Halo Victoria Ray at 24 weeks and 2 days old via cesarean because she was breached. Halo weighed 1 pound and 5 ounces and had to stay in the NICU at UAMS for 5 1/2 months. Coupled with Halo being a micro-preemie and Trisomy X syndrome, she is truly a miracle. I attribute the expertise of my doctors and the UAMS medical staff for ensuring my daughter and I both survived. Not everyone is blessed like I was to have doctors and a medical team that cared about our survival. I sincerely believe that Halo's story will spread faith, hope, light, love, and sunshine to the world.

Kafi Hunter

Testimonial

I first met Halo on 9/3/17 when she was almost 5 months old. She had spent those first five months of her life in the hospital. Now at home, she was on oxygen, thickened feedings and multiple medicines. She had grown from her birth weight of a little over one pound to 9# 3oz. Still, she could not do anything you would expect from a 5 month old, except smile. She was, and is, always engaging! To complicate her small size and early arrival, Halo has Trisomy X, which means having three X chromosomes instead of the usual two. Girls with this typically have developmental delays, some have autism, and the most severely affected have seizures and kidney disease. Adding that to her prematurity and small size, I will confess I had low expectations for her development. But only God knows what is in the future—so I joined with Halo's parents to give her every advantage. We were helped by the High Risk

Newborn clinic, the pulmonary clinic and the ENT clinic at Arkansas Children's Hospital.

Once Halo was medically stable, she enrolled at Pediatrics Plus in Little Rock and received speech, occupational and physical therapies. She made steady progress, and in fact graduated from intensive therapies. Her lungs have improved. She has problems with recurrent ear infections (she has very small ear canals with her Trisomy X) but these are improved with ear tubes. Amazingly, Halo has not required any hospital stays after graduating from the NICU!

Today, Halo is a normal sized 5 year old, who runs, talks and loves life. She will attend mainstream school and I expect great things from her. She and her parents are one of the joys of my practice.

Dr. Laura Williams

April 8, 2022

Triple X Syndrome

Overview

Triple X syndrome, also called trisomy X or 47,XXX, is a genetic disorder that affects about 1 in 1,000 females. Females normally have two X chromosomes in all cells — one X chromosome from each parent. In triple X syndrome, a female has three X chromosomes.

Many girls and women with triple X syndrome don't experience symptoms or have only mild symptoms. In others, symptoms may be more apparent — possibly including developmental delays and learning disabilities. Seizures and kidney problems occur in a small number of girls and women with triple X syndrome.

Treatment for triple X syndrome depends on which symptoms, if any, are present and their severity.

Symptoms

Signs and symptoms can vary greatly among girls and women with triple X syndrome. Many experience no noticeable effects or have only mild symptoms.

Being taller than average height is the most typical physical feature. Most females with triple X syndrome experience normal sexual development and have the ability to become pregnant. Some girls and women with triple X syndrome have intelligence in the normal range, but possibly slightly lower when compared with siblings. Others may have intellectual disabilities and sometimes may have behavioral problems.

Occasionally, significant symptoms may occur, which vary among individuals. These signs and symptoms may show up as:

- Delayed development of speech and language skills, as well as motor skills, such as sitting up and walking
- Learning disabilities, such as difficulty with reading, understanding or math
- Behavioral problems, such as attention-deficit/hyperactivity disorder (ADHD) or symptoms of autism spectrum disorder
- Psychological problems, such as anxiety and depression
- Problems with fine and gross motor skills, memory, judgment and information processing

Sometimes females with triple X syndrome have these signs and symptoms:

- Vertical folds of skin that cover the inner corners of the eyes (epicanthal folds)
- Widely spaced eyes
- Curved pinky fingers
- Flat feet
- Breastbone with an inward bowed shape
- Weak muscle tone (hypotonia)

Reference:

Mayo Foundation for Medical Education and Research. (n.d.). Triple X syndrome. Mayo Clinic. https://www.mayoclinic.org/diseases-conditions/triple-x-syndrome/symptoms-causes/syc-20350977

www.ingramcontent.com/pod-product-compliance
Lightning Source LLC
Chambersburg PA
CBHW080853060526
44107CB00129B/645